MAKE-UP LESSON P...
FOR ABSENT STUDE...

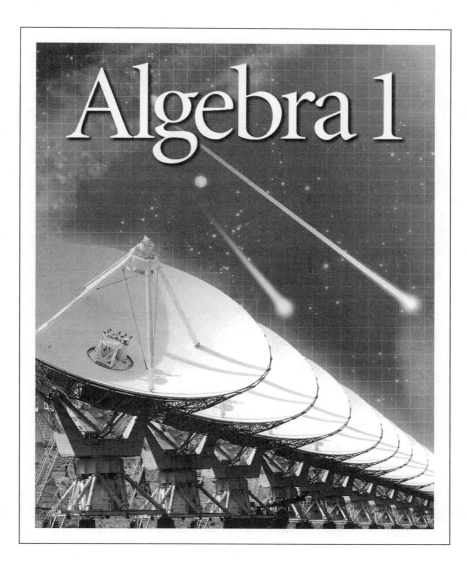

Algebra 1

HOLT, RINEHART AND WINSTON

A Harcourt Classroom Education Company

Austin • New York • Orlando • Atlanta • San Francisco • Boston • Dallas • Toronto • London

To the Student

Make-Up Lesson Planner for Absent Students is a set of quick, easy-to-use reference pages—one for each lesson in the *Pupil's Edition*—that provides the teacher with a checklist of assignments to give to the absent student. The checklist consists of key lesson goals, homework options, and additional homework assignments and resources. Space is provided for the teacher to further customize the assignment with additional comments and instructions.

Photo Credit
Front Cover: (background), Index Stock Photography Inc./Ron Russell; (bottom), Jean Miele MCMXCII/The Stock Market.

Printed in the United States of America

ISBN 0-03-064831-9

2   3   4   5   6   7   066   05   04   03   02

Table of Contents

Make-Up Lesson Planner for Absent Students
1.1 *Using Differences to Identify Patterns*

The items checked below were covered in class on _____ (date missed).

Goal 1: Use differences to identify patterns in number sequences.

_____ **Activity:** Exploring Differences

_____ Example 1: Find the terms of sequences by using constant differences.

_____ Example 2: Use constant differences to find maximum height.

Goal 2: Make predictions by using number sequence patterns.

_____ Example 3: Use problem-solving strategies to predict the number of conversations.

HOMEWORK

_____ Textbook exercises pages 8–10 (specify) _____

_____

_____ Look Back page 10   _____ Look Beyond page 10

_____ Portfolio Activity page 10   _____ Student Study Guide 1.1

ADDITIONAL HOMEWORK ASSIGNMENTS AND RESOURCES

_____ Practice 1.1   _____ Reteaching 1.1

_____ Student Technology Guide 1.1   _____ Lesson Quiz 1.1

_____ Internet Connection   GO TO: go.hrw.com   KEYWORD: MA1 College Costs

_____ Internet Connection   GO TO: go.hrw.com   KEYWORD: MA1 Triangular

Additional comments and instructions

_____

_____

_____

_____

_____

Make-Up Lesson Planner for Absent Students
1.2 *Variables, Expressions, and Equations*

The items checked below were covered in class on _____ (date missed).

Goal 1: Use variables, expressions, and equations to represent unknown quantities.

_____ Example 1: Make tables to show values of expressions given different values for variables.

_____ **Activity:** Modeling Data With a Table and an Equation

_____ Example 2: Make tables and write equations for given situations.

Goal 2: Represent real-world situations with equations and solve by guess-and-check.

_____ Example 3: Write and solve equations for real-world situations using guess-and-check strategies.

_____ Example 4: Use the constant feature of a calculator to solve problems, then write and solve equations using the table feature of graphics calculators.

HOMEWORK

_____ Textbook exercises pages 15–16 (specify) _____

_____

_____ Look Back page 17   _____ Look Beyond page 17

_____ Portfolio Activity page 17   _____ Student Study Guide 1.2

ADDITIONAL HOMEWORK ASSIGNMENTS AND RESOURCES

_____ Practice 1.2   _____ Reteaching 1.2

_____ Student Technology Guide 1.2   _____ Lesson Quiz 1.2

_____ Internet Connection   GO TO: go.hrw.com   KEYWORD: MA1 Mars Weight

Additional comments and instructions

_____

_____

_____

_____

NAME _____ CLASS _____ DATE _____

Make-Up Lesson Planner for Absent Students
1.3 The Algebraic Order of Operations

The items checked below were covered in class _____ (date missed).

Goal 1: Evaluate expressions using the algebraic order of operations.

_____ **Activity:** Exploring the Order of Operations

_____ Example 1: Use the algebraic order of operations to evaluate expressions.

_____ Example 2: Insert inclusion symbols to make equations true.

_____ Example 3: Use the algebraic order of operations to evaluate expressions given values of variables.

Goal 2: Use a calculator or computer to evaluate expressions with inclusion symbols.

_____ Example 4: Use graphics or scientific calculators to evaluate numerical expressions.

_____ Example 5: Evaluate expressions containing several pairs of inclusion symbols.

HOMEWORK

_____ Textbook exercises pages 21–23 (specify) _____

_____

_____ Look Back page 23   _____ Look Beyond page 23

_____ Student Study Guide 1.3

ADDITIONAL HOMEWORK ASSIGNMENTS AND RESOURCES

_____ Practice 1.3   _____ Reteaching 1.3

_____ Student Technology Guide 1.3   _____ Lesson Quiz 1.3

Additional comments and instructions

_____

_____

_____

_____

_____

Make-Up Lesson Planner for Absent Students
1.4 *Graphing With Coordinates*

The items checked below were covered in class on _____ (date missed).

Goal 1: Plot points and lines on a coordinate plane.

_____ Example 1: Determine quadrants and coordinates for points on graphs and plot given points.

Goal 2: Determine values for equations in the form $y = mx$ and graph ordered pairs on a coordinate plane.

_____ **Activity:** Graphing Data

_____ Example 2: Make a table for equations in the form $y = mx$, and find values for y by substituting values for x; graph ordered pairs, and connect the points with a line.

HOMEWORK

_____ Textbook exercises pages 27–28 (specify) _____

_____

_____ Look Back page 29   _____ Look Beyond page 29

_____ Portfolio Activity page 29   _____ Student Study Guide 1.4

ADDITIONAL HOMEWORK ASSIGNMENTS AND RESOURCES

_____ Practice 1.4   _____ Reteaching 1.4

_____ Student Technology Guide 1.4   _____ Lesson Quiz 1.4

_____ Internet Connection   GO TO: go.hrw.com   KEYWORD: MA1 Fibonacci

Additional comments and instructions

_____

_____

_____

_____

_____

_____

Make-Up Lesson Planner for Absent Students
1.5 Representing Linear Patterns

The items checked below were covered in class on _____ (date missed).

Goal 1: Generalize data patterns with equations.

_____ Example 1: Find first differences and write equations to represent data patterns.

Goal 2: Represent linear equations with graphs.

_____ **Activity:** Modeling an Equation With a Graph

_____ Example 2: Write and graph linear equations using dependent and independent variables.

_____ Example 3: Write and graph linear equations using dependent and independent variables.

HOMEWORK

_____ Textbook exercises pages 34–36 (specify) _____

_____

_____ Look Back page 36   _____ Look Beyond page 36

_____ Student Study Guide 1.5

ADDITIONAL HOMEWORK ASSIGNMENTS AND RESOURCES

_____ Practice 1.5   _____ Reteaching 1.5

_____ Student Technology Guide 1.5   _____ Lesson Quiz 1.5

Additional comments and instructions

_____
_____
_____
_____
_____
_____
_____
_____
_____

Make-Up Lesson Planner for Absent Students
1.6 Scatter Plots and Lines of Best Fit

The items checked below were covered in class on _____ (date missed).

Goal 1: Interpret data in a scatter plot.

_____ Example 1: Compare correlations in scatter plots to given statements.

Goal 2: Find a line of best fit on a scatter plot by inspection.

_____ **Activity:** Working With Trends in Data

_____ Example 2: Plot data, find lines of best fit, and describe correlations in scatter plots.

HOMEWORK

_____ Textbook exercises pages 40–42 (specify) _____

_____

_____ Look Back page 43   _____ Look Beyond page 43

_____ Portfolio Activity page 43   _____ Student Study Guide 1.6

ADDITIONAL HOMEWORK ASSIGNMENTS AND RESOURCES

_____ Practice 1.6   _____ Reteaching 1.6

_____ Student Technology Guide 1.6   _____ Lesson Quiz 1.6

_____ Internet Connection   GO TO: go.hrw.com   KEYWORD: MA1 Nose Count

Additional comments and instructions

_____

_____

_____

_____

_____

_____

_____

_____

_____

Make-Up Lesson Planner for Absent Students
2.1 *The Real Numbers and Absolute Value*

The items checked below were covered in class on _____ (date missed).

Goal 1: Compare real numbers.

_____ **Activity:** Terminating and Repeating Decimals

_____ Example 1: Insert ordering symbols to make statements true.

Goal 2: Simplify expressions involving opposites and absolute value.

_____ Example 2: Evaluate expressions involving opposites.

_____ Example 3: Find the absolute value of numbers.

HOMEWORK

_____ Textbook exercises pages 58–59 (specify) _____

_____

_____ Look Back page 59   _____ Look Beyond page 59

_____ Student Study Guide 2.1

ADDITIONAL HOMEWORK ASSIGNMENTS AND RESOURCES

_____ Practice 2.1   _____ Reteaching 2.1

_____ Student Technology Guide 2.1   _____ Lesson Quiz 2.1

Additional comments and instructions

_____

_____

_____

_____

_____

_____

_____

_____

_____

_____

NAME _____ CLASS _____ DATE _____

Make-Up Lesson Planner for Absent Students
2.2 *Adding Real Numbers*

The items checked below were covered in class on _____ (date missed).

Goal 1: Use algebra tiles to model addition.

_____ **Activity:** Modeling Addition—Unlike Signs

Goal 2: Add numbers with like and unlike signs.

_____ Example 1: Find sums of integers with like and unlike signs.

_____ Example 2: Find sums of decimals and fractions with like and unlike signs.

HOMEWORK

_____ Textbook exercises pages 64–65 (specify) _____

_____

_____ Look Back page 66   _____ Look Beyond page 66

_____ Portfolio Activity page 66   _____ Student Study Guide 2.2

ADDITIONAL HOMEWORK ASSIGNMENTS AND RESOURCES

_____ Practice 2.2   _____ Reteaching 2.2

_____ Student Technology Guide 2.2   _____ Lesson Quiz 2.2

Additional comments and instructions

_____

_____

_____

_____

_____

_____

_____

_____

_____

_____

Make-Up Lesson Planner for Absent Students
2.3 Subtracting Real Numbers

The items checked below were covered in class on _____ (date missed).

Goal 1: Use algebra tiles to model subtraction.

_____ Example 1: Use algebra tiles to model subtraction.

Goal 2: Determine subtraction in terms of addition.

_____ **Activity:** Modeling the Definition of Subtraction.

_____ Example 2: Use the definition of subtraction to find differences.

Goal 3: Subtract numbers with like and unlike signs.

_____ Example 3: Find distance between numbers on number lines.

HOMEWORK

_____ Textbook exercises pages 70–72 (specify) _____

_____

_____ Look Back page 72   _____ Look Beyond page 72

_____ Student Study Guide 2.3

ADDITIONAL HOMEWORK ASSIGNMENTS AND RESOURCES

_____ Practice 2.3   _____ Reteaching 2.3

_____ Student Technology Guide 2.3   _____ Lesson Quiz 2.3

_____ Internet Connection   GO TO: go.hrw.com   KEYWORD: MA1 U.S. Jobs

Additional comments and instructions

_____

_____

_____

_____

_____

_____

_____

Make-Up Lesson Planner for Absent Students
2.4 Multiplying and Dividing Real Numbers

The items checked below were covered in class on _____ (date missed).

Goal 1: Multiply and divide positive and negative numbers.

_____ **Activity:** Multiplication and Division Patterns

_____ Example 1: Find products of numbers with like and unlike signs.

_____ Example 2: Find quotients of numbers with like and unlike signs.

_____ Example 3: Find quotients by using reciprocals.

Goal 2: Define the Properties of Zero.

_____ Example 4: Find products or quotients by using the Properties of Zero.

HOMEWORK

_____ Textbook exercises pages 77–78 (specify) _____

_____

_____ Look Back page 79 _____ Look Beyond page 79

_____ Portfolio Activity page 79 _____ Student Study Guide 2.4

ADDITIONAL HOMEWORK ASSIGNMENTS AND RESOURCES

_____ Practice 2.4 _____ Reteaching 2.4

_____ Student Technology Guide 2.4 _____ Lesson Quiz 2.4

_____ Internet Connection   GO TO: go.hrw.com   KEYWORD: MA1 TEMP Extremes

Additional comments and instructions

_____

_____

_____

_____

_____

Make-Up Lesson Planner for Absent Students
2.5 Properties and Mental Computation

The items checked below were covered in class on _____ (date missed).

Goal 1: State and apply the Commutative, Associative, and Distributive Properties to perform mental computations.

_____ Example 1: Use the Commutative and Associative properties to compute mentally.

_____ **Activity:** Exploring Mental Computation

Goal 2: Apply the Commutative, Associative, and Distributive properties.

_____ Example 2: Use the Distributive Property to find the sum of two products.

_____ Example 3: Use the Distributive Property to find the opposite of expressions.

_____ Example 4: Use the Commutative, Associative, and Distributive Properties to prove mathematical statements, and justify each step.

_____ Example 5: Use the Commutative, Associative, and Distributive Properties to prove mathematical statements, and justify each step.

HOMEWORK

_____ Textbook exercises pages 85–87 (specify) _____

_____

_____ Look Back page 88   _____ Look Beyond page 88

_____ Portfolio Activity page 88   _____ Student Study Guide 2.5

ADDITIONAL HOMEWORK ASSIGNMENTS AND RESOURCES

_____ Practice 2.5   _____ Reteaching 2.5

_____ Student Technology Guide 2.5   _____ Lesson Quiz 2.5

_____ Internet Connection   GO TO: go.hrw.com   KEYWORD: MA1 Napier's Bones

_____ Internet Connection   GO TO: go.hrw.com   KEYWORD: MA1 Windchill

Additional comments and instructions

_____

_____

Make-Up Lesson Planner for Absent Students
2.6 Adding and Subtracting Expressions

The items checked below were covered in class on _____ (date missed).

Goal 1: Use the Distributive Property to combine like terms.

_____ **Activity:** Exploring Combining Like terms.

_____ Example 1: Use the Distributive Property to show true statements.

Goal 2: Simplify expressions with several variables.

_____ Example 2: Simplify addition and subtraction expressions by combining like terms.

HOMEWORK

_____ Textbook exercises pages 92–93 (specify) _____

_____

_____ Look Back page 93   _____ Look Beyond page 93

_____ Student Study Guide 2.6

ADDITIONAL HOMEWORK ASSIGNMENTS AND RESOURCES

_____ Practice 2.6   _____ Reteaching 2.6

_____ Student Technology Guide 2.6   _____ Lesson Quiz 2.6

_____ Internet Connection   GO TO: go.hrw.com   KEYWORD: MA1 Stocks

Additional comments and instructions

_____
_____
_____
_____
_____
_____
_____
_____
_____
_____

Make-Up Lesson Planner for Absent Students
2.7 *Multiplying and Dividing Expressions*

The items checked below were covered in class on _____ (date missed).

Goal 1: Multiply expressions containing variables.

_____ Example 1: Simplify expressions by using the definition of exponents and the Distributive Property.

_____ Example 2: Simplify expressions involving subtraction and multiplication by rewriting as addition problems.

Goal 2: Divide expressions containing variables.

_____ Example 3: Model division problems using algebra tiles.

_____ Example 4: Simplify expressions by dividing or rewriting as multiplication problems.

_____ Example 5: Write expressions to model situations using variables and division.

_____ Example 6: Simplify expressions by rewriting as multiplication problems.

HOMEWORK

_____ Textbook exercises pages 101–102 (specify) _____

_____

_____ Look Back page 103   _____ Look Beyond page 103

_____ Portfolio Activity page 103   _____ Student Study Guide 2.7

ADDITIONAL HOMEWORK ASSIGNMENTS AND RESOURCES

_____ Practice 2.7   _____ Reteaching 2.7

_____ Student Technology Guide 2.7   _____ Lesson Quiz 2.7

Additional comments and instructions

_____

_____

_____

_____

_____

Make-Up Lesson Planner for Absent Students
3.1 Solving Equations by Adding and Subtracting

The items checked below were covered in class on _____ (date missed).

Goal 1: Solve equations by using subtraction.

_____ Example 1: Solve equations using the Subtraction Property of Equality.

_____ Example 2: Write and solve equations by using the Subtraction Property of Equality.

Goal 2: Solve equations by using addition.

_____ **Activity:** Modeling the Addition Property of Equality

_____ Example 3: Solve equations by using the Addition Property of Equality.

_____ Example 4: Solve equations by using opposites and the Subtraction Property of Equality.

_____ Example 5: Write and solve equations involving addition or subtraction that represent real-world situations.

HOMEWORK

_____ Textbook exercises pages 118–120 (specify) _____

_____

_____ Look Back page 121   _____ Look Beyond page 121

_____ Portfolio Activity page 121   _____ Student Study Guide 3.1

ADDITIONAL HOMEWORK ASSIGNMENTS AND RESOURCES

_____ Practice 3.1   _____ Reteaching 3.1

_____ Student Technology Guide 3.1   _____ Lesson Quiz 3.1

_____ Internet Connection   GO TO: go.hrw.com   KEYWORD: MA1 Egyptian

Additional comments and instructions

_____

_____

_____

_____

Make-Up Lesson Planner for Absent Students
3.2 *Solving Equations by Multiplying and Dividing*

The items checked below were covered in class on _____ (date missed).

Goal 1: Solve equations by using division.

_____ **Activity:** Finding Solutions to Multiplication Equations

_____ Example 1: Solve multiplication equations by using the Division Property of Equality.

_____ Example 2: Solve geometry problems by writing multiplication equations and then using division.

Goal 2: Solve equations by using multiplication.

_____ Example 3: Solve division equations by using the Multiplication Property of Equality.

_____ Example 4: Solve equations by using reciprocals and the Multiplication Property of Equality.

_____ Example 5: Solve real-world problems by writing and solving division equations.

HOMEWORK

_____ Textbook exercises pages 126–127 (specify) _____

_____

_____ Look Back page 128   _____ Look Beyond page 128

_____ Portfolio Activity page 128   _____ Student Study Guide 3.2

ADDITIONAL HOMEWORK ASSIGNMENTS AND RESOURCES

_____ Practice 3.2   _____ Reteaching 3.2

_____ Student Technology Guide 3.2   _____ Lesson Quiz 3.2

_____ Internet Connection   GO TO: go.hrw.com   KEYWORD: MA1 Oscar at Bat

Additional comments and instructions

_____

_____

_____

Make-Up Lesson Planner for Absent Students
3.3 Solving Two-Step Equations

The items checked below were covered in class on _____ (date missed).

Goal 1: Write equations that represent real-world situations.

_____ **Activity:** Using a Table

_____ Example 1: Solve real-world problems by using two-step equations.

Goal 2: Solve two-step equations.

_____ Example 2: Solve two-step equations by using the Subtraction, Addition, and Division Properties of Equality.

_____ Example 3: Solve two-step equations by using the Subtraction, Addition, and Multiplication Properties of Equality.

_____ Example 4: Write and solve two-step equations that represent real-world situations.

HOMEWORK

_____ Textbook exercises pages 132–134 (specify) _____

_____

_____ Look Back page 134   _____ Look Beyond page 134

_____ Student Study Guide 3.3

ADDITIONAL HOMEWORK ASSIGNMENTS AND RESOURCES

_____ Practice 3.3   _____ Reteaching 3.3

_____ Student Technology Guide 3.3   _____ Lesson Quiz 3.3

Additional comments and instructions

_____

_____

_____

_____

_____

Make-Up Lesson Planner for Absent Students
3.4 Solving Multistep Equations

The items checked below were covered in class on _____ (date missed).

Goal: Write and solve multistep equations.

_____ **Activity:** Modeling Equations

_____ Example 1: Solve real-world problems by using multistep equations with the variable on each side of the equal sign.

_____ Example 2: Solve multistep equations with the variable on each side of the equal sign.

_____ Example 3: Solve multistep equations involving fractions.

_____ Example 4: Solve real-world problems by using multistep equations with the variable on each side of the equal sign.

HOMEWORK

_____ Textbook exercises pages 138–140 (specify) _____

_____

_____ Look Back page 140   _____ Look Beyond page 140

_____ Portfolio Activity page 140   _____ Student Study Guide 3.4

ADDITIONAL HOMEWORK ASSIGNMENTS AND RESOURCES

_____ Practice 3.4   _____ Reteaching 3.4

_____ Student Technology Guide 3.4   _____ Lesson Quiz 3.4

Additional comments and instructions

_____

_____

_____

_____

_____

_____

_____

Make-Up Lesson Planner for Absent Students
3.5 *Using the Distributive Property*

The items checked below were covered in class on _____ (date missed).

Goal 1: Solve real-world problems by using multistep equations.

_____ **Activity:** Using a Table to Solve for *x*

_____ Example 1: Solve real-world problems by using multistep equations, the Distributive Property, and other properties of equality.

Goal 2: Use the Distributive Property to solve equations.

_____ Example 2: Solve multistep equations using the Distributive Property.

_____ Example 3: Solve multistep equations where the solutions are special situations.

_____ Example 4: Solve multistep equations by using the properties of equality.

HOMEWORK

_____ Textbook exercises pages 144–146 (specify) _____

_____

_____ Look Back page 146   _____ Look Beyond page 146

_____ Student Study Guide 3.5

ADDITIONAL HOMEWORK ASSIGNMENTS AND RESOURCES

_____ Practice 3.5   _____ Reteaching 3.5

_____ Student Technology Guide 3.5   _____ Lesson Quiz 3.5

Additional comments and instructions

_____

_____

_____

_____

_____

_____

Make-Up Lesson Planner for Absent Students
3.6 Using Formulas and Literal Equations

The items checked below were covered in class on _____ (date missed).

Goal 1: Solve literal equations for a specific variable.

_____ Example 1: Use formulas by substituting specific variables.

_____ Example 2: Solve formulas for specific variables by using subtraction.

_____ Example 3: Solve formulas for specific variables by using multiplication.

_____ Example 4: Solve equations for indicated variables.

Goal 2: Use formulas to solve problems.

_____ Example 5: Use formulas to solve real-world problems involving the perimeter of a rectangle.

_____ Example 6: Use formulas to solve real-world problems involving the area of a trapezoid.

_____ Example 7: Use the circumference formula to solve for the radius of a circle.

HOMEWORK

_____ Textbook exercises pages 151–152 (specify) _____

_____

_____ Look Back page 153   _____ Look Beyond page 153

_____ Portfolio Activity page 153   _____ Student Study Guide 3.6

ADDITIONAL HOMEWORK ASSIGNMENTS AND RESOURCES

_____ Practice 3.6   _____ Reteaching 3.6

_____ Student Technology Guide 3.6 _____ Lesson Quiz 3.6

Additional comments and instructions

_____

_____

_____

_____

Make-Up Lesson Planner for Absent Students
4.1 *Using Proportional Reasoning*

The items checked below were covered in class on _____ (date missed).

Goal 1: Identify the means and extremes of a proportion.

_____ **Activity:** Exploring Relationships in Proportions

_____ Example 1: Use cross products to determine whether proportions are true.

_____ Example 2: Use cross products to find values of variables that make proportions true.

Goal 2: Use proportions to solve problems.

_____ Example 3: Use proportions to solve a problem involving similar polygons.

_____ Example 4: Make tables, write proportions, and use cross products to solve real-world problems.

HOMEWORK

_____ Textbook exercises pages 168–169 (specify) _____

_____

_____ Look Back page 170   _____ Look Beyond page 170

_____ Portfolio Activity page 170   _____ Student Study Guide 4.1

ADDITIONAL HOMEWORK ASSIGNMENTS AND RESOURCES

_____ Practice 4.1   _____ Reteaching 4.1

_____ Student Technology Guide 4.1   _____ Lesson Quiz 4.1

Additional comments and instructions

_____

_____

_____

_____

_____

_____

Make-Up Lesson Planner for Absent Students
4.2 Percent Problems

The items checked below were covered in class on _____ (date missed).

Goal 1: Find equivalent fractions, decimals, ands percents.

_____ Example 1: Write percents as decimals, fractions, or mixed numbers in simplest terms.

Goal 2: Solve problems involving percent.

_____ Example 2: Solve real-world problems involving percent by using the proportion method and the equation method.

_____ Example 3: Model and solve real-world problems involving percent discounts.

_____ Example 4: Use proportions to solve problems involving percent sales tax.

_____ Example 5: Solve problems involving percent of increase.

_____ Example 6: Solve problems involving percent of decrease.

HOMEWORK

_____ Textbook exercises pages 175–177 (specify) _____

_____

_____ Look Back page 177   _____ Look Beyond page 177

_____ Student Study Guide 4.2

ADDITIONAL HOMEWORK ASSIGNMENTS AND RESOURCES

_____ Practice 4.2   _____ Reteaching 4.2

_____ Student Technology Guide 4.2   _____ Lesson Quiz 4.2

_____ Internet Connection   GO TO: go.hrw.com   KEYWORD: MA1 Rural Exodus

Additional comments and instructions

_____

_____

_____

_____

Make-Up Lesson Planner for Absent Students
4.3 *Introduction to Probability*

The items checked below were covered in class on _____ (date missed).

Goal: Find the experimental probability that an event will occur.

_____ **Activity:** Two Probability Experiments

_____ Example 1: Find experimental probabilities based on number cube experiments.

_____ Example 2: Find experimental probabilities based on coin experiments.

HOMEWORK

_____ Textbook exercises pages 183–184 (specify) _____

_____

_____ Look Back page 185   _____ Look Beyond page 185

_____ Portfolio Activity page 185   _____ Student Study Guide 4.3

ADDITIONAL HOMEWORK ASSIGNMENTS AND RESOURCES

_____ Practice 4.3   _____ Reteaching 4.3

_____ Student Technology Guide 4.3   _____ Lesson Quiz 4.3

_____ Internet Connection   GO TO: go.hrw.com   KEYWORD: MA1 Simpson's Paradox

_____ Internet Connection   GO TO: go.hrw.com   KEYWORD: MA1 Coin Toss

Additional comments and instructions

_____

_____

_____

_____

_____

_____

_____

_____

_____

Make-Up Lesson Planner for Absent Students
4.4 Measures of Central Tendency

The items checked below were covered in class on _____ (date missed).

Goal 1: Find the mean, median, mode, and range of a data set.

_____ Example 1: Find the mean, median, mode, and range of real-world data sets involving batting averages.

_____ Example 2: Find the mean, median, mode, and range of data sets.

_____ Example 3: Solve problems involving test score averages by using the formula for mean.

Goal 2: Represent data with frequency tables.

_____ **Activity:** Using Frequency Tables

_____ Example 4: Find the mean, median, mode, and range of a real-world problem involving swimming statistics presented in a frequency table.

HOMEWORK

_____ Textbook exercises pages 190–192 (specify) _____

_____

_____ Look Back page 192   _____ Look Beyond page 192

_____ Student Study Guide 4.4

ADDITIONAL HOMEWORK ASSIGNMENTS AND RESOURCES

_____ Practice 4.4   _____ Reteaching 4.4

_____ Student Technology Guide 4.4   _____ Lesson Quiz 4.4

_____ Internet Connection   GO TO: go.hrw.com   KEYWORD: MA1 Voting Schemes

Additional comments and instructions

_____

_____

_____

_____

_____

Make-Up Lesson Planner for Absent Students
4.5 *Graphing Data*

The items checked below were covered in class on _____ (date missed).

Goal 1: Analyze graphs in order to find misleading presentation of data.

_____ **Activity:**   Misleading Graphs

Goal 2: Interpret line graphs, bar graphs, and circle graphs.

_____ Example 1: Interpret line graphs and construct bar graphs.

_____ Example 2: Interpret circle graphs.

Goal 3: Represent data with circle graphs.

_____ Example 3: Use information given in tables to create circle graphs.

HOMEWORK

_____ Textbook exercises pages 198–200 (specify) _____

_____

_____ Look Back page 200   _____ Look Beyond page 200

_____ Student Study Guide 4.5

ADDITIONAL HOMEWORK ASSIGNMENTS AND RESOURCES

_____ Practice 4.5   _____ Reteaching 4.5

_____ Student Technology Guide 4.5   _____ Lesson Quiz 4.5

Additional comments and instructions

_____

_____

_____

_____

_____

_____

_____

_____

Make-Up Lesson Planner for Absent Students
4.6 Other Data Displays

The items checked below were covered in class on _____ (date missed).

Goal: Interpret and construct stem-and-leaf plots, histograms, and box-and-whisker plots.

_____ Example 1: Construct stem-and-leaf plots from given data.

_____ **Activity:** Interpreting a Box-and-Whisker Plot

_____ Example 2: Use stem-and-leaf plots to construct box-and-whisker-plots.

HOMEWORK

_____ Textbook exercises pages 205–206 (specify) _____

_____

_____ Look Back page 207   _____ Look Beyond page 207

_____ Portfolio Activity page 207   _____ Student Study Guide 4.6

ADDITIONAL HOMEWORK ASSIGNMENTS AND RESOURCES

_____ Practice 4.6   _____ Reteaching 4.6

_____ Student Technology Guide 4.6   _____ Lesson Quiz 4.6

_____ Internet Connection   GO TO: go.hrw.com   KEYWORD: MA1 Endangered Species

Additional comments and instructions

_____

_____

_____

_____

_____

_____

_____

_____

_____

_____

Make-Up Lesson Planner for Absent Students
5.1 *Linear Functions and Graphs*

The items checked below were covered in class on _____ (date missed).

Goal 1: Determine whether a relation is a function.

_____ **Activity:** Exploring Relations and Functions

Goal 2: Describe the domain and range of a function.

_____ **Example 1:** Determine whether relations are functions and describe the domain and range.

_____ **Example 2:** Complete ordered pairs so that the ordered pairs are solutions to given equations containing variables x and y.

_____ **Example 3:** Write equations and ordered pairs for functions that represent real-world situations, and determine the domain and range.

_____ **Example 4:** Identify dependent and independent variables, describe the domain and range, and write equations from given graphs of linear functions that represent real-world situations.

HOMEWORK

_____ Textbook exercises pages 223–225 (specify) _____

_____

_____ Look Back page 225   _____ Look Beyond page 225

_____ Student Study Guide 5.1

ADDITIONAL HOMEWORK ASSIGNMENTS AND RESOURCES

_____ Practice 5.1   _____ Reteaching 5.1

_____ Student Technology Guide 5.1   _____ Lesson Quiz 5.1

Additional comments and instructions

_____

_____

_____

_____

Make-Up Lesson Planner for Absent Students
5.2 Defining Slope

The items checked below were covered in class on _____ (date missed).

Goal 1: Calculate the slope of a line by using the rise and the run.

_____ Example 1: Calculate the slope of a line that represents a real-world situation by finding the ratio of rise to run.

Goal 2: Calculate the slope of a line from the ratio of the differences in x-and y-coordinates.

_____ Example 2: Calculate the slope of a line by using the x- and y-intercepts.

_____ Example 3: Calculate the slope of a line by using coordinates of given points.

_____ **Activity:** Fitting a Line to a Point

_____ Example 4: Calculate the slope of a horizontal line by using points on the line.

_____ Example 5: Calculate the slope of a vertical line by using points on the line.

HOMEWORK

_____ Textbook exercises pages 230–233 (specify) _____

_____

_____ Look Back page 233   _____ Look Beyond page 233

_____ Portfolio Activity page 233   _____ Student Study Guide 5.2

ADDITIONAL HOMEWORK ASSIGNMENTS AND RESOURCES

_____ Practice 5.2   _____ Reteaching 5.2

_____ Student Technology Guide 5.2   _____ Lesson Quiz 5.2

_____ Internet Connection   GO TO: go.hrw.com   KEYWORD: MA1 Marathon

_____ Internet Connection   GO TO: go.hrw.com   KEYWORD: MA1 Golden Ratio

Additional comments and instructions

_____

_____

NAME _____ CLASS _____ DATE _____

Make-Up Lesson Planner for Absent Students
5.3 Rate of Change and Direct Variation

The items checked below were covered in class on _____ (date missed).

Goal 1: Find the rate of change from a graph.

_____ Example 1: Use the rate of change to find the speed of a car from a graph.

_____ Example 2: Identify rates of change from graphed data.

Goal 2: Solve and graph direct-variation equations.

_____ Example 3: Find constants of variation and write direct-variation equations.

_____ Example 4: Solve direct-variation problems by using proportions.

_____ Example 5: Use Hooke's Law to find the force, constant, or distance of a spring given part of the information.

_____ **Activity:** Exploring the Graphs of Direct Variations

HOMEWORK

_____ Textbook exercises pages 240–242 (specify) _____

_____

_____ Look Back pages 242–243   _____ Look Beyond page 243

_____ Portfolio Activity page 243   _____ Student Study Guide 5.3

ADDITIONAL HOMEWORK ASSIGNMENTS AND RESOURCES

_____ Practice 5.3   _____ Reteaching 5.3

_____ Student Technology Guide 5.3   _____ Lesson Quiz 5.3

_____ Internet Connection   GO TO: go.hrw.com   KEYWORD: MA1 Import Export

_____ Internet Connection   GO TO: go.hrw.com   KEYWORD: MA1 Ordered Pairs

Additional comments and instructions

_____

_____

_____

_____

Make-Up Lesson Planner for Absent Students
5.4 *The Slope-Intercept Form*

The items checked below were covered in class on _____ (date missed).

Goal 1: Define and explain the components of the slope-intercept form of a linear equation.

_____ Example 1: Explain and use the slope-intercept form to construct graphs of equations that represent real-world situations.

_____ **Activity:** Exploring Graphs of Linear Functions

Goal 2: Use the slope-intercept form of a linear equation.

_____ Example 2: Write equations in slope-intercept form to model real-world situations.

_____ Example 3: Identify the *x*- and *y*-intercepts of lines given the slope-intercept form of linear equations.

_____ Example 4: Write equations for horizontal and vertical lines using the *x*- or *y*-intercepts.

HOMEWORK

_____ Textbook exercises pages 249–251 (specify) _____

_____

_____ Look Back page 251   _____ Look Beyond page 251

_____ Student Study Guide 5.4

ADDITIONAL HOMEWORK ASSIGNMENTS AND RESOURCES

_____ Practice 5.4   _____ Reteaching 5.4

_____ Student Technology Guide 5.4   _____ Lesson Quiz 5.4

Additional comments and instructions

_____

_____

_____

_____

_____

_____

Make-Up Lesson Planner for Absent Students
5.5 *The Standard and Point-Slope Forms*

The items checked below were covered in class on _____ (date missed).

Goal 1: Define and use the standard form of a linear equation.

_____ Example 1: Write linear equations in standard form.

_____ Example 2: Write linear equations in standard form to model real-world situations, and graph the equations.

Goal 2: Define and use the point-slope form of a linear equation.

_____ Example 3: Write equations in point-slope form given a point on the line and the slope of the line.

_____ Example 4: Write linear equations in point-slope form and standard form.

_____ Example 5: Write equations in point-slope form given two points on the line.

HOMEWORK

_____ Textbook exercises pages 256–257 (specify) _____

_____

_____ Look Back page 257   _____ Look Beyond page 257

_____ Student Study Guide 5.5

ADDITIONAL HOMEWORK ASSIGNMENTS AND RESOURCES

_____ Practice 5.5   _____ Reteaching 5.5

_____ Student Technology Guide 5.5   _____ Lesson Quiz 5.5

_____ Internet Connection   GO TO: go.hrw.com   KEYWORD: MA1 Check Digits

Additional comments and instructions

_____

_____

_____

_____

_____

_____

Make-Up Lesson Planner for Absent Students
5.6 Parallel and Perpendicular Lines

The items checked below were covered in class on _____ (date missed).

Goal: Identify parallel and perpendicular lines by comparing slopes, and write equations of lines that are parallel and perpendicular to given lines.

_____ Example 1: Write equations in slope-intercept form of lines parallel to given lines.

_____ **Activity:** Exploring Slopes and Perpendicular Lines

_____ Example 2: Write equations in slope-intercept form of lines perpendicular to given lines.

_____ Example 3: Write equation in point-slope form of lines perpendicular to given lines.

HOMEWORK

_____ Textbook exercises pages 261–262 (specify) _____

_____

_____ Look Back page 263   _____ Look Beyond page 263

_____ Portfolio Activity page 263   _____ Student Study Guide 5.6

ADDITIONAL HOMEWORK ASSIGNMENTS AND RESOURCES

_____ Practice 5.6   _____ Reteaching 5.6

_____ Student Technology Guide 5.6 _____ Lesson Quiz 5.6

Additional comments and instructions

_____

_____

_____

_____

_____

_____

_____

_____

Make-Up Lesson Planner for Absent Students
6.1 *Solving Inequalities*

The items checked below were covered in class on _____ (date missed).

Goal 1: State and use symbols of inequality.

_____ Example 1: Use symbols of inequality to write inequalities that model real-world situations.

Goal 2: Solve inequalities that involve addition and subtraction.

_____ Example 2: Solve inequalities by using the Addition Property of Equality.

_____ Example 3: Graph solutions to inequalities.

_____ Example 4: Write inequalities that describe points graphed on number lines.

HOMEWORK

_____ Textbook exercises pages 279–281 (specify) _____

_____

_____ Look Back page 281   _____ Look Beyond page 281

_____ Portfolio Activity page 281   _____ Student Study Guide 6.1

ADDITIONAL HOMEWORK ASSIGNMENTS AND RESOURCES

_____ Practice 6.1   _____ Reteaching 6.1

_____ Student Technology Guide 6.1   _____ Lesson Quiz 6.1

Additional comments and instructions

_____
_____
_____
_____
_____
_____
_____
_____

Make-Up Lesson Planner for Absent Students
6.2 *Multistep Inequalities*

The items checked below were covered in class on _____ (date missed).

Goal 1: State and apply the Multiplication and Division Properties of Inequality.

_____ **Activity:** Multiplying and Dividing Inequalities

_____ Example 1: Use multiplication and division of positive numbers to solve inequalities.

_____ Example 2: Use multiplication and division of negative numbers to solve inequalities.

Goal 2: Solve multistep inequalities in one variable.

_____ Example 3: Use properties of equality to solve multistep inequalities in one variable.

_____ Example 4: Write and solve inequalities that represent real-world situations.

HOMEWORK

_____ Textbook exercises pages 286–287 (specify) _____

_____

_____ Look Back page 288   _____ Look Beyond page 288

_____ Portfolio Activity page 288   _____ Student Study Guide 6.2

ADDITIONAL HOMEWORK ASSIGNMENTS AND RESOURCES

_____ Practice 6.2   _____ Reteaching 6.2

_____ Student Technology Guide 6.2   _____ Lesson Quiz 6.2

_____ Internet Connection   GO TO: go.hrw.com   KEYWORD: MA1 Puzzles

Additional comments and instructions

_____

_____

_____

_____

_____

Make-Up Lesson Planner for Absent Students
6.3 Compound Inequalities

The items checked below were covered in class on _____ (date missed).

Goal 1: Graph the solution sets of compound inequalities.

_____ Example 1: Write compound inequalities and graph the solution sets to show conjunctions that model real-world situations.

_____ Example 2: Write compound inequalities and graph the solution sets to show disjunctions that model real-world situations.

Goal 2: Solve compound inequalities.

_____ Example 3: Graph given compound inequalities.

_____ Example 4: Solve and graph given compound inequalities that are conjunctions.

_____ Example 5: Solve and graph given compound inequalities that are disjunctions.

_____ Example 6: Solve and graph given compound inequalities where the solution set is all real numbers.

HOMEWORK

_____ Textbook exercises pages 292–293 (specify) _____

_____

_____ Look Back page 293   _____ Look Beyond page 293

_____ Student Study Guide 6.3

ADDITIONAL HOMEWORK ASSIGNMENTS AND RESOURCES

_____ Practice 6.3   _____ Reteaching 6.3

_____ Student Technology Guide 6.3 _____ Lesson Quiz 6.3

_____ Internet Connection   GO TO: go.hrw.com   KEYWORD: MA1 Hurricane Force

Additional comments and instructions

_____

_____

_____

Make-Up Lesson Planner for Absent Students
6.4 Absolute-Value Functions

The items checked below were covered in class on _____ (date missed).

Goal 1: Find absolute values, and graph absolute-value functions.

_____ Example 1: Find absolute values.

_____ Example 2: Find the domain and range of absolute-value functions, and graph absolute-value functions for given values.

Goal 2: Explore basic transformations of the absolute-value function.

_____ **Activity:** Manipulating the Absolute-Value Function

_____ Example 3: Graph the absolute-value function with given transformation functions, and identify the type of transformation.

HOMEWORK

_____ Textbook exercises pages 297–299 (specify) _____

_____

_____ Look Back page 299   _____ Look Beyond page 299

_____ Student Study Guide 6.4

ADDITIONAL HOMEWORK ASSIGNMENTS AND RESOURCES

_____ Practice 6.4   _____ Reteaching 6.4

_____ Student Technology Guide 6.4   _____ Lesson Quiz 6.4

Additional comments and instructions

_____

_____

_____

_____

_____

_____

_____

_____

Make-Up Lesson Planner for Absent Students
6.5 Absolute-Value Equations and Inequalities

The items checked below were covered in class on _____ (date missed).

Goal 1: Solve absolute-value equations.

_____ Example 1: Solve absolute-value equations that model real-world situations.

_____ Example 2: Solve absolute-value equations.

Goal 2: Solve absolute-value inequalities, and express the solution as a range of values on a number line.

_____ Example 3: Solve absolute-value inequalities, and express the solution as a range of values.

_____ Example 4: Solve absolute-value inequalities, where the solution is a conjunction, and express the solution as a range of values on a number line.

_____ Example 5: Solve absolute-value inequalities, where the solution is a disjunction, and express the solution as a range of values on a number line.

_____ **Activity:** Graphing to Find Solutions

_____ Example 6: Use graphing to solve absolute-value equalities and inequalities.

HOMEWORK

_____ Textbook exercises pages 305–307 (specify) _____

_____

_____ Look Back page 307 _____ Look Beyond page 307

_____ Student Study Guide 6.5

ADDITIONAL HOMEWORK ASSIGNMENTS AND RESOURCES

_____ Practice 6.5 _____ Reteaching 6.5

_____ Student Technology Guide 6.5 _____ Lesson Quiz 6.5

Additional comments and instructions

_____

_____

Make-Up Lesson Planner for Absent Students
7.1 Graphing Systems of Equations

The items checked below were covered in class on _____ (date missed).

Goal: Graph systems of linear equations, and solve by inspecting the graphs for solutions.

_____ Example 1: Solve given systems of equations by graphing.

_____ Example 2: Change equations in a given system to slope-intercept form, and solve by graphing.

_____ **Activity:** Exploring Approximate Solutions

_____ Example 3: Find approximate solutions to systems of equations by graphing.

_____ Example 4: Write systems of linear equations to model real-world problems, and solve the systems by graphing.

HOMEWORK

_____ Textbook exercises pages 323–325 (specify) _____

_____

_____ Look Back page 325   _____ Look Beyond page 325

_____ Portfolio Activity page 325   _____ Student Study Guide 7.1

ADDITIONAL HOMEWORK ASSIGNMENTS AND RESOURCES

_____ Practice 7.1   _____ Reteaching 7.1

_____ Student Technology Guide 7.1   _____ Lesson Quiz 7.1

_____ Internet Connection   GO TO: go.hrw.com   KEYWORD: MA1 Prices

_____ Internet Connection   GO TO: go.hrw.com   KEYWORD: MA1 Athletes

Additional comments and instructions

_____

_____

_____

_____

Make-Up Lesson Planner for Absent Students
7.2 *The Substitution Method*

The items checked below were covered in class on _____ (date missed).

Goal: Find exact solutions to systems of linear equations by using the substitution method.

_____ **Activity:** Exploring Substitution

_____ Example 1: Solve given systems of two linear equations by substituting the known value of a variable.

_____ Example 2: Solve given systems of two linear equations by substituting the known expression for a variable.

_____ Example 3: Solve given systems of two linear equations by finding and substituting an expression for a variable.

_____ Example 4: Solve real-world problems by using systems of linear equations and the substitution method.

HOMEWORK

_____ Textbook exercises pages 329–330 (specify) _____

_____

_____ Look Back page 330   _____ Look Beyond page 330

_____ Student Study Guide 7.2

ADDITIONAL HOMEWORK ASSIGNMENTS AND RESOURCES

_____ Practice 7.2   _____ Reteaching 7.2

_____ Student Technology Guide 7.2   _____ Lesson Quiz 7.2

_____ Internet Connection   GO TO: go.hrw.com   KEYWORD: MA1 House Gender

Additional comments and instructions

_____

_____

_____

_____

_____

Make-Up Lesson Planner for Absent Students
7.3 *The Elimination Method*

The items checked below were covered in class on _____ (date missed).

Goal: Use the elimination method to solve a system of equations.

_____ **Activity:** Exploring Using Opposites

_____ Example 1: Use the elimination method to solve a system of equations containing opposites.

_____ Example 2: Use the elimination method to solve a system of equations by multiplying first to produce a system containing opposites.

_____ Example 3: Use the elimination method to solve a system of equations that represents a real-world situation.

HOMEWORK

_____ Textbook exercises pages 335–336 (specify) _____

_____

_____ Look Back page 337   _____ Look Beyond page 337

_____ Portfolio Activity page 337   _____ Student Study Guide 7.3

ADDITIONAL HOMEWORK ASSIGNMENTS AND RESOURCES

_____ Practice 7.3   _____ Reteaching 7.3

_____ Student Technology Guide 7.3   _____ Lesson Quiz 7.3

_____ Internet Connection   GO TO: go.hrw.com   KEYWORD: MA1 Sports Opera

Additional comments and instructions

_____

_____

_____

_____

_____

_____

_____

_____

Make-Up Lesson Planner for Absent Students
7.4 Consistent and Inconsistent Systems

The items checked below were covered in class on _____ (date missed).

Goal 1: Identify consistent and inconsistent systems of equations.

_____ **Activity:** Exploring No Solution and Many Solutions

_____ Example 1: Determine consistent and inconsistent systems of equations by solving equations in slope-intercept form algebraically and graphically.

_____ Example 2: Determine consistent and inconsistent systems of equations by solving equations in standard form algebraically and graphically.

_____ Example 3: Plot data and write systems of equations to solve real-world problems involving inconsistent systems.

Goal 2: Identify dependent and independent systems of equations.

_____ Example 4: Determine whether systems are dependent or independent by solving algebraically and graphically.

HOMEWORK

_____ Textbook exercises pages 342–344 (specify) _____

_____

_____ Look Back page 344   _____ Look Beyond page 344

_____ Student Study Guide 7.4

ADDITIONAL HOMEWORK ASSIGNMENTS AND RESOURCES

_____ Practice 7.4   _____ Reteaching 7.4

_____ Student Technology Guide 7.4   _____ Lesson Quiz 7.4

Additional comments and instructions

_____

_____

_____

_____

Make-Up Lesson Planner for Absent Students
7.5 *Systems of Inequalities*

The items checked below were covered in class on _____ (date missed).

Goal 1: Graph the solution to a linear inequality.

_____ **Activity:** Modeling a Linear Inequality With a Graph

_____ Example 1: Graph linear inequalities in two variables.

_____ Example 2: Graph linear inequalities in two variables.

Goal 2: Graph the solution to a system of linear inequalities.

_____ Example 3: Solve systems of linear inequalities by graphing.

_____ Example 4: Use graphing to solve systems of linear inequalities that model real-world situations.

HOMEWORK

_____ Textbook exercises pages 350–351 (specify) _____

_____

_____ Look Back page 352   _____ Look Beyond page 352

_____ Portfolio Activity page 352   _____ Student Study Guide 7.5

ADDITIONAL HOMEWORK ASSIGNMENTS AND RESOURCES

_____ Practice 7.5   _____ Reteaching 7.5

_____ Student Technology Guide 7.5   _____ Lesson Quiz 7.5

Additional comments and instructions

_____

_____

_____

_____

_____

_____

_____

_____

_____

Make-Up Lesson Planner for Absent Students
7.6 *Classic Puzzles in Two Variables*

The items checked below were covered in class on _____ (date missed).

Goal: Solve traditional puzzles in two variables.

_____ Example 1: Solve traditional age puzzles.

_____ Example 2: Solve traditional wind and current puzzles.

_____ Example 3: Solve traditional number-digit puzzles.

_____ Example 4: Solve traditional coin puzzles.

_____ Example 5: Solve traditional chemical-solution puzzles.

HOMEWORK

_____ Textbook exercises pages 357–359 (specify) _____

_____

_____ Look Back page 359   _____ Look Beyond page 359

_____ Student Study Guide 7.6

ADDITIONAL HOMEWORK ASSIGNMENTS AND RESOURCES

_____ Practice 7.6   _____ Reteaching 7.6

_____ Student Technology Guide 7.6   _____ Lesson Quiz 7.6

Additional comments and instructions

_____
_____
_____
_____
_____
_____
_____
_____
_____
_____

Make-Up Lesson Planner for Absent Students
8.1 *Laws of Exponents: Multiplying Monomials*

The items checked below were covered in class on _____ (date missed).

Goal 1: Define exponents and powers.

_____ **Example 1:** Evaluate numbers raised to exponents.

_____ **Activity:** Multiplying Powers

Goal 2: Find products of powers.

_____ **Example 2:** Simplify powers of products containing the same base.

_____ **Example 3:** Solve real-world problems involving products of powers.

Goal 3: Simplify products of monomials.

_____ **Example 4:** Simplify products of monomials containing exponents.

_____ **Example 5:** Use expressions containing exponents to model real-world situations involving volume.

HOMEWORK

_____ Textbook exercises pages 374–375 (specify) _____

_____

_____ Look Back page 376   _____ Look Beyond page 376

_____ Portfolio Activity page 376   _____ Student Study Guide 8.1

ADDITIONAL HOMEWORK ASSIGNMENTS AND RESOURCES

_____ Practice 8.1   _____ Reteaching 8.1

_____ Student Technology Guide 8.1 _____ Lesson Quiz 8.1

_____ Internet Connection   GO TO: go.hrw.com   KEYWORD: MA1 Number Sieves

_____ Internet Connection   GO TO: go.hrw.com   KEYWORD: Doubling

Additional comments and instructions

_____

_____

Make-Up Lesson Planner for Absent Students

8.2 *Laws of Exponents: Powers and Products*

The items checked below were covered in class on _____ (date missed).

Goal 1: Find the power of a power.

_____ **Activity:** Raising a Power to a Power

_____ Example 1: Simplify expressions containing powers raised to powers.

_____ Example 2: Simplify monomials containing exponents raised to a power by regrouping.

Goal 2: Find the power of a product.

_____ Example 3: Verify the Power-of-a-Product Property by finding values of expressions using two different methods.

_____ Example 4: Simplify monomials containing exponents raised to a power by using the Power-of-a-Product Property.

_____ Example 5: Compare volumes by using the volume formula of a sphere.

_____ Example 6: Simplify monomials containing exponents by using even and odd powers of -1.

HOMEWORK

_____ Textbook exercises pages 381–382 (specify) _____

_____

_____ Look Back page 382   _____ Look Beyond page 382

_____ Student Study Guide 8.2

ADDITIONAL HOMEWORK ASSIGNMENTS AND RESOURCES

_____ Practice 8.2   _____ Reteaching 8.2

_____ Student Technology Guide 8.2   _____ Lesson Quiz 8.2

Additional comments and instructions

_____

_____

Make-Up Lesson Planner for Absent Students

8.3 *Laws of Exponents: Dividing Monomials*

The items checked below were covered in class on _____ (date missed).

Goal 1: Simplify quotients of powers.

_____ **Activity:** Discovering a Rule for Quotients of Powers

_____ Example 1: Simplify quotients of powers with like bases.

_____ Example 2: Simplify quotients of powers with like bases and variable exponents.

_____ Example 3: Simplify quotients of powers containing like and unlike bases.

_____ Example 4: Solve a geometry problem involving comparing volumes and surface area of spheres.

Goal 2: Simplify powers of fractions.

_____ Example 5: Simplify expressions containing powers of fractions.

HOMEWORK

_____ Textbook exercises pages 387−389 (specify) _____

_____

_____ Look Back page 389   _____ Look Beyond page 389

_____ Student Study Guide 8.3

ADDITIONAL HOMEWORK ASSIGNMENTS AND RESOURCES

_____ Practice 8.3   _____ Reteaching 8.3

_____ Student Technology Guide 8.3   _____ Lesson Quiz 8.3

Additional comments and instructions

_____

_____

_____

_____

_____

Make-Up Lesson Planner for Absent Students
8.4 Negative and Zero Exponents

The items checked below were covered in class on _____ (date missed).

Goal: Simplify expressions containing negative and zero exponents.

_____ **Example 1:** Evaluate products and quotients of numbers containing negative and positive exponents.

_____ **Activity:** Defining x^0

_____ **Example 2:** Simplify and write products and quotients with only positive exponents.

_____ **Example 3:** Simplify and write expressions with only positive exponents.

HOMEWORK

_____ Textbook exercises pages 393–394 (specify) _____

_____

_____ Look Back page 395   _____ Look Beyond page 395

_____ Portfolio Activity page 395   _____ Student Study Guide 8.4

ADDITIONAL HOMEWORK ASSIGNMENTS AND RESOURCES

_____ Practice 8.4   _____ Reteaching 8.4

_____ Student Technology Guide 8.4   _____ Lesson Quiz 8.4

_____ Internet Connection   GO TO: go.hrw.com   KEYWORD: MA1 Centenarians

Additional comments and instructions

_____
_____
_____
_____
_____
_____
_____
_____
_____
_____

Make-Up Lesson Planner for Absent Students
8.5 *Scientific Notation*

The items checked below were covered in class on _____ (date missed).

Goal 1: Recognize the need for special notation in scientific calculations.

_____ **Activity:** Exploring Powers of 10

_____ Example 1: Write very large numbers from astronomy in scientific notation.

Goal 2: Perform computations involving scientific notation.

_____ Example 2: Use the properties of exponents and scientific notation to perform calculations.

_____ Example 3: Write very small numbers from physics in scientific notation.

_____ Example 4: Express products and quotients using scientific notation.

_____ Example 5: Solve problems written in scientific notation with scientific calculators

HOMEWORK

_____ Textbook exercises pages 401–403 (specify) _____

_____

_____ Look Back page 403   _____ Look Beyond page 403

_____ Student Study Guide 8.5

ADDITIONAL HOMEWORK ASSIGNMENTS AND RESOURCES

_____ Practice 8.5   _____ Reteaching 8.5

_____ Student Technology Guide 8.5   _____ Lesson Quiz 8.5

_____ Internet Connection   GO TO: go.hrw.com   KEYWORD: MA1 Scientific Notation

Additional comments and instructions

_____

_____

_____

Make-Up Lesson Planner for Absent Students
8.6 Exponential Functions

The items checked below were covered in class on _____ (date missed).

Goal 1: Understand exponential functions and how they are used.

_____ **Activity:** Population Growth

_____ Example 1: Solve a problem about exponential population growth by using a scientific calculator.

_____ Example 2: Make tables of values and graph exponential functions.

Goal 2: Recognize differences between graphs of exponential functions with different bases.

_____ Example 3: Graph different exponential functions on the same axes to see differences and similarities.

HOMEWORK

_____ Textbook exercises pages 407–408 (specify) _____

_____

_____ Look Back page 408   _____ Look Beyond page 408

_____ Student Study Guide 8.6

ADDITIONAL HOMEWORK ASSIGNMENTS AND RESOURCES

_____ Practice 8.6   _____ Reteaching 8.6

_____ Student Technology Guide 8.6   _____ Lesson Quiz 8.6

_____ Internet Connection   GO TO: go.hrw.com   KEYWORD: MA1 Snowflake

Additional comments and instructions

_____

_____

_____

_____

_____

_____

Make-Up Lesson Planner for Absent Students
8.7 Applications of Exponential Functions

The items checked below were covered in class on _____ (date missed).

Goal: Use exponential functions to model applications that include growth and decay in different contexts.

_____ **Example 1:** Solve a problem about carbon-14 dating to estimate the age of an object.

_____ **Example 2:** Use exponential functions to perform financial calculations involving interest rates.

_____ **Example 3:** Solve growth rate problems by using the general growth formula.

_____ **Example 4:** Use exponential functions to solve problems involving percentage rates.

_____ **Example 5:** Use graphs and tables of exponential functions to solve problems involving rates of growth or decay.

HOMEWORK

_____ Textbook exercises pages 413–414 (specify) _____

_____

_____ Look Back page 415   _____ Look Beyond page 415

_____ Portfolio Activity page 415   _____ Student Study Guide 8.7

ADDITIONAL HOMEWORK ASSIGNMENTS AND RESOURCES

_____ Practice 8.7   _____ Reteaching 8.7

_____ Student Technology Guide 8.7   _____ Lesson Quiz 8.7

_____ Internet Connection   GO TO: go.hrw.com   KEYWORD: MA1 Compound Interest

Additional comments and instructions

_____

_____

_____

_____

_____

Make-Up Lesson Planner for Absent Students
9.1 *Adding and Subtracting Polynomials*

The items checked below were covered in class on _____ (date missed).

Goal: Add and subtract polynomials.

_____ **Example 1:** Write polynomials in standard form.

_____ **Activity:** Polynomial Addition and Subtraction

_____ **Example 2:** Find sums of polynomials by using the vertical and horizontal forms.

_____ **Example 3:** Write polynomial expressions for perimeters by using the vertical and horizontal forms.

_____ **Example 4:** Find differences of polynomials by using the vertical and horizontal forms.

HOMEWORK

_____ Textbook exercises pages 429–431 (specify) _____

_____

_____ Look Back page 431   _____ Look Beyond page 431

_____ Student Study Guide 9.1

ADDITIONAL HOMEWORK ASSIGNMENTS AND RESOURCES

_____ Practice 9.1   _____ Reteaching 9.1

_____ Student Technology Guide 9.1   _____ Lesson Quiz 9.1

_____ Internet Connection   GO TO: go.hrw.com   KEYWORD: MA1 Poly Ops

Additional comments and instructions

_____

_____

_____

_____

_____

_____

_____

_____

Make-Up Lesson Planner for Absent Students
9.2 Modeling Polynomial Multiplication

The items checked below were covered in class on _____ (date missed).

Goal 1: Use algebra tiles to model the products of binomials.

_____ **Activity:** Exploring Multiplication of Polynomials

_____ Example 1: Use algebra tiles to find products of binomials.

Goal 2: Mentally simplify special products of binomials.

_____ Example 2: Use the rules for special products to find products of binomials.

HOMEWORK

_____ Textbook exercises pages 435–436 (specify) _____

_____

_____ Look Back page 437   _____ Look Beyond page 437

_____ Portfolio Activity page 437   _____ Student Study Guide 9.2

ADDITIONAL HOMEWORK ASSIGNMENTS AND RESOURCES

_____ Practice 9.2   _____ Reteaching 9.2

_____ Student Technology Guide 9.2 _____ Lesson Quiz 9.2

_____ Internet Connection   GO TO: go.hrw.com   KEYWORD: MA1 Pascal

Additional comments and instructions

_____

_____

_____

_____

_____

_____

_____

_____

_____

Make-Up Lesson Planner for Absent Students
9.3 Multiplying Binomials

The items checked below were covered in class on _____ (date missed).

Goal 1: Find products of binomials by using the Distributive Property.

_____ Example 1: Use the Distributive Property to show the multi-plication of binomials.

_____ **Activity:** The Distributive Property and Algebra Tiles

Goal 2: Find products of binomials by using the FOIL method.

_____ Example 2: Use the FOIL method to find products of binomials.

_____ Example 3: Find area by using given measurements from a diagram and operations on binomials including multiplication.

HOMEWORK

_____ Textbook exercises pages 441–442 (specify) _____

_____

_____ Look Back page 442   _____ Look Beyond page 442

_____ Student Study Guide 9.3

ADDITIONAL HOMEWORK ASSIGNMENTS AND RESOURCES

_____ Practice 9.3   _____ Reteaching 9.3

_____ Student Technology Guide 9.3   _____ Lesson Quiz 9.3

Additional comments and instructions

_____

_____

_____

_____

_____

_____

_____

_____

Make-Up Lesson Planner for Absent Students
9.4 Polynomial Functions

The items checked below were covered in class on _____ (date missed).

Goal: Solve problems involving polynomial functions.

_____ **Example 1:** Write and evaluate polynomial functions to find volumes.

_____ **Activity:** Exploring Volume and Surface Area

_____ **Example 2:** Use polynomial functions to find volumes, and find dimensions associated with given percent increases in volume.

_____ **Example 3:** Show that polynomial functions are equivalent for specific integer values.

HOMEWORK

_____ Textbook exercises pages 446–447 (specify) _____

_____

_____ Look Back page 447   _____ Look Beyond page 447

_____ Portfolio Activity page 447   _____ Student Study Guide 9.4

ADDITIONAL HOMEWORK ASSIGNMENTS AND RESOURCES

_____ Practice 9.4   _____ Reteaching 9.4

_____ Student Technology Guide 9.4   _____ Lesson Quiz 9.4

_____ Internet Connection   GO TO: go.hrw.com   KEYWORD: MA1 Regression

_____ Internet Connection   GO TO: go.hrw.com   KEYWORD: MA1 Figurate

Additional comments and instructions

_____

_____

_____

_____

_____

_____

_____

Make-Up Lesson Planner for Absent Students
9.5 *Common Factors*

The items checked below were covered in class on _____ (date missed).

Goal 1: Factor a polynomial by using the greatest common factor.

_____ Example 1: Factor binomials by using greatest common factors.

Goal 2: Factor polynomials by using a binomial factor.

_____ Example 2: Factor binomials by using common binomial factors and the Distributive Property.

_____ Example 3: Factor polynomials by grouping.

HOMEWORK

_____ Textbook exercises pages 450–451 (specify) _____

_____

_____ Look Back page 451   _____ Look Beyond page 451

_____ Student Study Guide 9.5

ADDITIONAL HOMEWORK ASSIGNMENTS AND RESOURCES

_____ Practice 9.5   _____ Reteaching 9.5

_____ Student Technology Guide 9.5   _____ Lesson Quiz 9.5

Additional comments and instructions

_____
_____
_____
_____
_____
_____
_____
_____
_____
_____

Make-Up Lesson Planner for Absent Students
9.6 *Factoring Special Polynomials*

The items checked below were covered in class on _____ (date missed).

Goal 1: Factor perfect-square trinomials.

_____ Example 1: Identify expressions as perfect-square trinomials.

_____ Example 2: Factor perfect-square trinomials.

Goal 2: Factor the difference of two squares.

_____ Example 3: Identify expressions as the difference of two squares.

_____ **Activity:** Patterns in Differences of Two Squares

_____ Example 4: Factor differences of two squares.

_____ Example 5: Find a numerical product by using the difference of two squares.

HOMEWORK

_____ Textbook exercises pages 455–456 (specify) _____

_____

_____ Look Back page 457   _____ Look Beyond page 457

_____ Portfolio Activity page 457   _____ Student Study Guide 9.6

ADDITIONAL HOMEWORK ASSIGNMENTS AND RESOURCES

_____ Practice 9.6   _____ Reteaching 9.6

_____ Student Technology Guide 9.6 _____ Lesson Quiz 9.6

Additional comments and instructions

_____

_____

_____

_____

_____

_____

_____

Make-Up Lesson Planner for Absent Students
9.7 Factoring Quadratic Trinomials

The items checked below were covered in class on _____ (date missed).

Goal: Factor quadratic trinomials.

_____   Example 1:   Factor quadratic trinomials by using the FOIL method when the constant term is positive and the coefficient of the middle term is positive.

_____   Example 2:   Factor quadratic trinomials by using the FOIL method when the constant term is positive and the coefficient of the middle term is negative.

_____   Example 3:   Factor quadratic polynomials when the constant term of the trinomial is negative and the coefficient of the middle term is positive.

_____   Example 4:   Factor quadratic polynomials when the constant term of the trinomial is negative and the coefficient of the middle term is negative.

_____   Example 5:   Factor out common monomials in order to write trinomials as products.

HOMEWORK

_____   Textbook exercises pages 462–463 (specify) _____

_____

_____   Look Back page 463   _____   Look Beyond page 463

_____   Student Study Guide 9.7

ADDITIONAL HOMEWORK ASSIGNMENTS AND RESOURCES

_____   Practice 9.7   _____   Reteaching 9.7

_____   Student Technology Guide 9.7   _____   Lesson Quiz 9.7

Additional comments and instructions

_____

_____

_____

_____

_____

Make-Up Lesson Planner for Absent Students
9.8 *Solving Equations by Factoring*

The items checked below were covered in class on _____ (date missed).

Goal: Solve polynomial equations by factoring.

_____ **Activity:** Exploring the Zeros of a Function

_____ Example 1: Find the zeros of polynomial functions.

_____ Example 2: Solve polynomial equations by factoring and using the Zero Product Property.

_____ Example 3: Solve polynomial equations by factoring and using the Zero Product Property.

HOMEWORK

_____ Textbook exercises pages 467–468 (specify) _____

_____

_____ Look Back page 469   _____ Look Beyond page 469

_____ Portfolio Activity page 469   _____ Student Study Guide 9.8

ADDITIONAL HOMEWORK ASSIGNMENTS AND RESOURCES

_____ Practice 9.8   _____ Reteaching 9.8

_____ Student Technology Guide 9.8   _____ Lesson Quiz 9.8

_____ Internet Connection   GO TO: go.hrw.com   KEYWORD: MA1 Zeros

_____ Internet Connection   GO TO: go.hrw.com   KEYWORD: MA1 Rectangular

Additional comments and instructions

_____

_____

_____

_____

_____

_____

_____

_____

Make-Up Lesson Planner for Absent Students
10.1 *Graphing Parabolas*

The items checked below were covered in class on _____ (date missed).

Goal 1: Discover how adding a constant to the parent function $y = x^2$ affects the graph of the function.

_____ **Activity:** Transformations of $y = x^2$

_____ Example 1: Identify the vertices and axes of symmetry for graphs of quadratic equations.

Goal 2: Use the zeros of a quadratic function to find the vertex of the graph of the function.

_____ Example 2: Use zeros of quadratic functions to find the vertices of their graphs.

_____ Example 3: Solve physics problems involving free fall by using quadratic functions.

HOMEWORK

_____ Textbook exercises pages 484–485 (specify) _____

_____

_____ Look Back page 485   _____ Look Beyond page 485

_____ Student Study Guide 10.1

ADDITIONAL HOMEWORK ASSIGNMENTS AND RESOURCES

_____ Practice 10.1   _____ Reteaching 10.1

_____ Student Technology Guide 10.1   _____ Lesson Quiz 10.1

_____ Internet Connection   GO TO: go.hrw.com   KEYWORD: MA1 Parabolas

Additional comments and instructions

_____

_____

_____

_____

_____

_____

Make-Up Lesson Planner for Absent Students
10.2 *Solving Equations by Using Square Roots*

The items checked below were covered in class on _____ (date missed).

Goal 1: Solve equations of the form $ax^2 = k$.

_____ **Activity:** Tables for Falling Objects

_____ Example 1: Solve equations of the form $x^2 = k$.

_____ Example 2: Solve equations of the form $ax^2 = k$ to find times for falling objects to fall.

Goal 2: Solve equations of the form $x^2 = k$ where x is replaced by an algebraic expression.

_____ Example 3: Solve equations of the form $x^2 = k$, where x is replaced by an algebraic expression.

HOMEWORK

_____ Textbook exercises pages 489–490 (specify) _____

_____

_____ Look Back page 491   _____ Look Beyond page 491

_____ Portfolio Activity page 491   _____ Student Study Guide 10.2

ADDITIONAL HOMEWORK ASSIGNMENTS AND RESOURCES

_____ Practice 10.2   _____ Reteaching 10.2

_____ Student Technology Guide 10.2 _____ Lesson Quiz 10.2

Additional comments and instructions

_____

_____

_____

_____

_____

_____

_____

_____

Make-Up Lesson Planner for Absent Students
10.3 *Completing the Square*

The items checked below were covered in class on _____ (date missed).

Goal 1: Form a perfect-square trinomial from a given quadratic binomial.

_____ **Activity:** Completing the Square With Tiles

_____ Example 1: Complete the squares by finding values that make perfect-square trinomials when the coefficients of the x-terms are positive.

_____ Example 2: Complete the squares by finding values that make perfect-square trinomials when the coefficients of the x-terms are negative.

Goal 2: Write a given quadratic function in vertex form.

_____ Example 3: Rewrite given quadratic functions in vertex form, and identify the vertex of its graph.

HOMEWORK

_____ Textbook exercises pages 495–497 (specify) _____

_____

_____ Look Back page 497   _____ Look Beyond page 497

_____ Portfolio Activity page 497   _____ Student Study Guide 10.3

ADDITIONAL HOMEWORK ASSIGNMENTS AND RESOURCES

_____ Practice 10.3   _____ Reteaching 10.3

_____ Student Technology Guide 10.3   _____ Lesson Quiz 10.3

_____ Internet Connection   GO TO: go.hrw.com   KEYWORD: MA1 Gravity

Additional comments and instructions

_____

_____

_____

_____

_____

NAME _____ CLASS _____ DATE _____

Make-Up Lesson Planner for Absent Students
10.4 Solving Equations of the Form $x^2 + bx + c = 0$

The items checked below were covered in class on _____ (date missed).

Goal: Solve quadratic equations by completing the square.

_____ Example 1: Find zeros of quadratic functions.

_____ Example 2: Solve quadratic functions by completing the square and by factoring.

_____ Example 3: Find the value of x when y is a given number in a quadratic equation by graphing and by factoring and using the Zero Product Property.

_____ **Activity:** Package Design

_____ Example 4: Find the points where graphs of quadratic and linear functions intersect by graphing and by using substitution to solve a system.

HOMEWORK

_____ Textbook exercises pages 502–503 (specify) _____

_____

_____ Look Back page 503   _____ Look Beyond page 503

_____ Student Study Guide 10.4

ADDITIONAL HOMEWORK ASSIGNMENTS AND RESOURCES

_____ Practice 10.4   _____ Reteaching 10.4

_____ Student Technology Guide 10.4   _____ Lesson Quiz 10.4

_____ Internet Connection   GO TO: go.hrw.com   KEYWORD: MA1 Motion Equations

Additional comments and instructions

_____

_____

_____

_____

_____

_____

Make-Up Lesson Planner for Absent Students
10.5 The Quadratic Formula

The items checked below were covered in class on _____ (date missed).

Goal 1: Use the quadratic formula to find solutions to quadratic equations.

_____ **Example 1:** Use the quadratic formula to solve quadratic equations.

Goal 2: Use the quadratic formula to find the zeros of quadratic functions.

_____ **Example 2:** Use the quadratic formula to find the zeros of quadratic functions.

Goal 3: Evaluate discriminants to determine how many real roots a quadratic equation has and whether it can be factored.

_____ **Example 3:** Evaluate discriminants to determine the number of real solutions to quadratic equations.

_____ **Example 4:** Evaluate discriminants of quadratic equations to determine whether the equations can be factored.

HOMEWORK

_____ Textbook exercises pages 509–510 (specify) _____

_____

_____ Look Back page 510   _____ Look Beyond page 510

_____ Student Study Guide 10.5

ADDITIONAL HOMEWORK ASSIGNMENTS AND RESOURCES

_____ Practice 10.5   _____ Reteaching 10.5

_____ Student Technology Guide 10.5   _____ Lesson Quiz 10.5

_____ Internet Connection   GO TO: go.hrw.com   KEYWORD: MA1 Quadratic Formula

Additional comments and instructions

_____

_____

_____

Make-Up Lesson Planner for Absent Students
10.6 *Graphing Quadratic Inequalities*

The items checked below were covered in class on _____ (date missed).

Goal: Solve and graph quadratic inequalities, and test solution regions.

_____ **Activity:** Production and Profit

_____ Example 1: Solve and graph quadratic inequalities in one variable, and test solution regions.

_____ Example 2: Graph quadratic inequalities in two variables, and test solution regions.

_____ Example 3: Determine which inequalities satisfy given situations.

HOMEWORK

_____ Textbook exercises pages 514–515 (specify) _____

_____

_____ Look Back page 515   _____ Look Beyond page 515

_____ Portfolio Activity page 515 _____ Student Study Guide 10.6

ADDITIONAL HOMEWORK ASSIGNMENTS AND RESOURCES

_____ Practice 10.6   _____ Reteaching 10.6

_____ Student Technology Guide 10.6 _____ Lesson Quiz 10.6

Additional comments and instructions

_____
_____
_____
_____
_____
_____
_____
_____

Make-Up Lesson Planner for Absent Students
11.1 *Inverse Variation*

The items checked below were covered in class on _____ (date missed).

Goal: Define and use two different forms of inverse variation to study real-world situations.

_____ **Activity:** Exploring Inverse Variation

_____ Example 1: Write inverse-variation equations that satisfy given conditions.

_____ Example 2: Write and graph inverse-variation equations that represent experimental physics data.

_____ Example 3: Solve and graph inverse-variation equations derived by using the Rule of 72.

HOMEWORK

_____ Textbook exercises pages 529–531 (specify) _____

_____

_____ Look Back page 531   _____ Look Beyond page 531

_____ Student Study Guide 11.1

ADDITIONAL HOMEWORK ASSIGNMENTS AND RESOURCES

_____ Practice 11.1   _____ Reteaching 11.1

_____ Student Technology Guide 11.1   _____ Lesson Quiz 11.1

_____ Internet Connection   GO TO: go.hrw.com   KEYWORD: MA1 Gear Ratios

Additional comments and instructions

_____

_____

_____

_____

_____

_____

_____

_____

Make-Up Lesson Planner for Absent Students
11.2 *Rational Expressions and Functions*

The items checked below were covered in class on _____ (date missed).

Goal: Define and illustrate the use of rational expressions and functions.

_____ **Activity:** Graphing a Rational Function

_____ Example 1: Find the domain of rational functions.

_____ Example 2: Rewrite rational functions in simplest terms, and use their graphs to solve problems.

HOMEWORK

_____ Textbook exercises pages 535–536 (specify) _____

_____

_____ Look Back page 537   _____ Look Beyond page 537

_____ Student Study Guide 11.2

ADDITIONAL HOMEWORK ASSIGNMENTS AND RESOURCES

_____ Practice 11.2   _____ Reteaching 11.2

_____ Student Technology Guide 11.2   _____ Lesson Quiz 11.2

Additional comments and instructions

_____

_____

_____

_____

_____

_____

_____

_____

_____

_____

_____

Make-Up Lesson Planner for Absent Students
11.3 *Simplifying Rational Expressions*

The items checked below were covered in class on _____ (date missed).

Goal 1: Factor to simplify rational expressions, and state the restrictions on the variable of a simplified rational expression.

_____ **Activity:** Exploring Rational Functions

_____ Example 1: Simplify rational expressions by factoring, and state any restrictions on the variable.

_____ Example 2: Simplify rational expressions by factoring, and state any restrictions on the variable.

_____ Example 3: Simplify rational expressions by factoring, and state any restrictions on the variable.

Goal 2: Extend simplification techniques to other algebraic fractions.

_____ Example 4: Use simplification techniques to find ratios of volumes to surface area.

HOMEWORK

_____ Textbook exercises pages 541–542 (specify) _____
_____

_____ Look Back page 543   _____ Look Beyond page 543

_____ Portfolio Activity page 543   _____ Student Study Guide 11.3

ADDITIONAL HOMEWORK ASSIGNMENTS AND RESOURCES

_____ Practice 11.3   _____ Reteaching 11.3

_____ Student Technology Guide 11.3   _____ Lesson Quiz 11.3

_____ Internet Connection   GO TO: go.hrw.com   KEYWORD: MA1 Lever

Additional comments and instructions

_____
_____
_____
_____
_____
_____

Make-Up Lesson Planner for Absent Students
11.4 Operations With Rational Expressions

The items checked below were covered in class on _____ (date missed).

Goal: Add, subtract, multiply and divide rational expressions.

_____ Example 1: Multiply two rational expressions, and state any restrictions on the variable.

_____ Example 2: Multiply three rational expressions, and state any restrictions on the variable.

_____ Example 3: Divide rational expressions, and state any restrictions on the variable.

_____ Example 4: Add rational expressions, and state any restrictions on the variable.

_____ Example 5: Subtract rational expressions, and state any restrictions on the variable.

HOMEWORK

_____ Textbook exercises pages 549–551 (specify) _____

_____

_____ Look Back page 551   _____ Look Beyond page 551

_____ Student Study Guide 11.4

ADDITIONAL HOMEWORK ASSIGNMENTS AND RESOURCES

_____ Practice 11.4   _____ Reteaching 11.4

_____ Student Technology Guide 11.4   _____ Lesson Quiz 11.4

_____ Internet Connection   GO TO: go.hrw.com   KEYWORD: MA1 Coaster Math

Additional comments and instructions

_____

_____

_____

_____

_____

_____

Make-Up Lesson Planner for Absent Students
11.5 Solving Rational Equations

The items checked below were covered in class on _____ (date missed).

Goal: Solve rational equations by using the common-denominator method and by graphing.

_____ **Activity:** Guessing and Checking to Solve a Problem

_____ Example 1: Solve rational equations by using the common-denominator method and by using the graphing method.

_____ Example 2: Write rational equations to model real-world situations, and solve by using the common-denominator method.

HOMEWORK

_____ Textbook exercises pages 555–557 (specify) _____

_____

_____ Look Back page 557   _____ Look Beyond page 557

_____ Portfolio Activity page 557   _____ Student Study Guide 11.5

ADDITIONAL HOMEWORK ASSIGNMENTS AND RESOURCES

_____ Practice 11.5   _____ Reteaching 11.5

_____ Student Technology Guide 11.5   _____ Lesson Quiz 11.5

_____ Internet Connection   GO TO: go.hrw.com   KEYWORD: MA1 Rational Functions

_____ Internet Connection   GO TO: go.hrw.com   KEYWORD: MA1 Gears

Additional comments and instructions

_____

_____

_____

_____

_____

_____

_____

Make-Up Lesson Planner for Absent Students
11.6 *Proof in Algebra*

The items checked below were covered in class on _____ (date missed).

Goal 1: Define the parts of a conditional statement.

_____ **Activity:** Inductive Reasoning

Goal 2: Prove theorems stated in conditional form.

_____ Example 1: Prove theorems stated in conditional form by using algebraic properties.

_____ Example 2: Prove theorems stated in conditional form by using the definition of odd numbers.

_____ Example 3: Prove theorems stated in conditional form by using the definition of even numbers.

_____ Example 4: Prove conjectures by using the definitions of odd and even numbers.

_____ Example 5: Tell whether given statements are always true, sometimes true, or never true.

_____ Example 6: Determine whether sets are closed under specified operations.

HOMEWORK

_____ Textbook exercises pages 563–565 (specify) _____

_____

_____ Look Back page 565   _____ Look Beyond page 565

_____ Student Study Guide 11.6

ADDITIONAL HOMEWORK ASSIGNMENTS AND RESOURCES

_____ Practice 11.6   _____ Reteaching 11.6

_____ Student Technology Guide 11.6   _____ Lesson Quiz 11.6

Additional comments and instructions

_____

_____

_____

Make-Up Lesson Planner for Absent Students
12.1 *Operations With Radicals*

The items checked below were covered in class on _____ (date missed).

Goal 1: Identify or estimate square roots.

_____ Example 1: Evaluate square roots.

Goal 2: Perform mathematical operations with radicals, and write the results in simplest radical form.

_____ Example 2: Simplify radical expressions involving sums and differences.

_____ **Activity:** Operations and Radical Expressions

_____ Example 3: Write square roots in simplest radical form.

_____ Example 4: Simplify expressions involving products of square roots.

_____ Example 5: Simplify quotients of square roots.

HOMEWORK

_____ Textbook exercises pages 580–582 (specify) _____

_____

_____ Look Back page 582   _____ Look Beyond page 582

_____ Student Study Guide 12.1

ADDITIONAL HOMEWORK ASSIGNMENTS AND RESOURCES

_____ Practice 12.1   _____ Reteaching 12.1

_____ Student Technology Guide 12.1   _____ Lesson Quiz 12.1

_____ Internet Connection   GO TO: go.hrw.com   KEYWORD: MA1 Continued Fractions

Additional comments and instructions

_____

_____

_____

_____

_____

Make-Up Lesson Planner for Absent Students
12.2 *Square-Root Functions and Radical Equations*

The items checked below were covered in class on _____ (date missed)

Goal 1: Solve equations containing radicals.

_____ Example 1: Use formulas containing radicals.

_____ Example 2: Solve radical equations with a variable on one side of the equation.

_____ Example 3: Solve radical equations with a variable on each side of the equation.

Goal 2: Solve equations by using radicals.

_____ Example 4: Solve equations containing squared variables by using radicals.

_____ Example 5: Solve area problems by using radicals.

_____ Example 6: Solve equations containing perfect-square trinomials by factoring and using radicals.

HOMEWORK

_____ Textbook exercises pages 588–589 (specify) _____

_____

_____ Look Back page 590   _____ Look Beyond page 590

_____ Portfolio Activity page 590   _____ Student Study Guide 12.2

ADDITIONAL HOMEWORK ASSIGNMENTS AND RESOURCES

_____ Practice 12.2   _____ Reteaching 12.2

_____ Student Technology Guide 12.2   _____ Lesson Quiz 12.2

_____ Internet Connection   GO TO: go.hrw.com   KEYWORD: MA1 Planets

Additional comments and instructions

_____

_____

_____

_____

_____

Make-Up Lesson Planner for Absent Students
12.3 *The Pythagorean Theorem*

The items checked below were covered in class on _____ (date missed).

Goal 1: Find a side length of a right triangle given the lengths of its other two sides.

_____ **Activity:** Exploring Right Triangles

_____ Example 1: Find a side length of a right triangle given the lengths of its two other sides.

Goal 2: Apply the Pythagorean Theorem to real-world problems.

_____ Example 2: Use the Pythagorean Theorem to solve a real-world problem involving distance.

_____ Example 3: Use the Pythagorean Theorem to solve a geometry problem involving the dimensions of a pyramid.

HOMEWORK

_____ Textbook exercises pages 595–596 (specify) _____

_____

_____ Look Back page 597   _____ Look Beyond page 597

_____ Student Study Guide 12.3

ADDITIONAL HOMEWORK ASSIGNMENTS AND RESOURCES

_____ Practice 12.3   _____ Reteaching 12.3

_____ Student Technology Guide 12.3   _____ Lesson Quiz 12.3

Additional comments and instructions

_____

_____

_____

_____

_____

_____

_____

Make-Up Lesson Planner for Absent Students
12.4 The Distance Formula

The items checked below were covered in class on _____ (date missed).

Goal 1: Use the distance formula to find the distance between two points in a coordinate plane.

_____ Example 1: Find distances between pairs of points by using the distance formula.

Goal 2: Determine whether a triangle is a right triangle.

_____ Example 2: Given the coordinates of their vertices, determine whether triangles are right triangles.

Goal 3: Apply the midpoint formula.

_____ **Activity:** Deriving the Midpoint Formula

_____ Example 3: Use a coordinate plane to apply the midpoint formula to solve a real-world problem.

_____ Example 4: Use the midpoint formula to solve geometry problems involving circles and their centers and diameters.

HOMEWORK

_____ Textbook exercises pages 602–604 (specify) _____

_____

_____ Look Back page 605   _____ Look Beyond page 605

_____ Portfolio Activity page 605   _____ Student Study Guide 12.4

ADDITIONAL HOMEWORK ASSIGNMENTS AND RESOURCES

_____ Practice 12.4   _____ Reteaching 12.4

_____ Student Technology Guide 12.4   _____ Lesson Quiz 12.4

_____ Internet Connection   GO TO: go.hrw.com   KEYWORD: MA1 Theodorus

Additional comments and instructions

_____

_____

_____

NAME _____ CLASS _____ DATE _____

Make-Up Lesson Planner for Absent Students
12.5 Geometric Properties

The items checked below were covered in class on _____ (date missed).

Goal 1: Define and use the equation of a circle.

_____ Example 1: Use the distance formula to find equations of circles.

Goal 2: Use the coordinate plane to investigate the diagonals of a rectangle and the midsegment of a triangle.

_____ **Activity:** The Diagonals of a Rectangle

_____ Example 2: Test the Triangle Midsegment Theorem for a given triangle on a coordinate plane.

_____ Example 3: Write equations for real-world problems by using the distance formula.

HOMEWORK

_____ Textbook exercises pages 610–612 (specify) _____

_____

_____ Look Back page 612   _____ Look Beyond page 612

_____ Student Study Guide 12.5

ADDITIONAL HOMEWORK ASSIGNMENTS AND RESOURCES

_____ Practice 12.5   _____ Reteaching 12.5

_____ Student Technology Guide 12.5   _____ Lesson Quiz 12.5

Additional comments and instructions

_____

_____

_____

_____

_____

_____

_____

_____

Make-Up Lesson Planner for Absent Students
12.6 The Tangent Function

The items checked below were covered in class on _____ (date missed).

Goal 1: Identify and use the tangent ratio in a right triangle.

_____ **Activity:** The Tangent Ratio in Similar Triangles

_____ Example 1: Find the tangent ratios of acute angles in right triangles. Then find the measures of the acute angles.

Goal 2: Find unknown side and angle measures in right triangles.

_____ Example 2: Solve real-world problems by using tangent ratios to find angle measures in right triangles.

_____ Example 3: Solve real-world problems by using tangent ratios to find the measure of unknown sides in right triangles.

HOMEWORK

_____ Textbook exercises pages 618–620 (specify) _____

_____

_____ Look Back page 620   _____ Look Beyond page 620

_____ Student Study Guide 12.6

ADDITIONAL HOMEWORK ASSIGNMENTS AND RESOURCES

_____ Practice 12.6   _____ Reteaching 12.6

_____ Student Technology Guide 12.6   _____ Lesson Quiz 12.6

_____ Internet Connection   GO TO: go.hrw.com   KEYWORD: MA1 Right Triangles

Additional comments and instructions

_____

_____

_____

_____

_____

_____

Make-Up Lesson Planner for Absent Students
12.7 The Sine and Cosine Functions

The items checked below were covered in class on _____ (date missed).

Goal: Find unknown side and angle measures in right triangles.

_____   **Activity:**   Exploring Sines and Cosines

_____   Example 1:   Use sine ratios to find the missing sides in right triangles.

_____   Example 2:   Use cosine ratios to find the missing angle measures in right triangles.

HOMEWORK

_____   Textbook exercises pages 624–626 (specify) _____

_____

_____   Look Back page 626   _____   Look Beyond page 627

_____   Portfolio Activity page 627   _____   Student Study Guide 12.7

ADDITIONAL HOMEWORK ASSIGNMENTS AND RESOURCES

_____   Practice 12.7   _____   Reteaching 12.7

_____   Student Technology Guide 12.7 _____   Lesson Quiz 12.7

_____   Internet Connection   GO TO: go.hrw.com   KEYWORD: MA1 Trig Functions

_____   Internet Connection   GO TO: go.hrw.com   KEYWORD: MA1 Golden

Additional comments and instructions

_____

_____

_____

_____

_____

_____

_____

_____

_____

_____

Make-Up Lesson Planner for Absent Students
12.8 *Introduction to Matrices*

The items checked below were covered in class on _____ (date missed).

Goal 1: Determine whether two matrices are equal.

_____ Example 1: Determine whether two matrices are equal.

Goal 2: Add, subtract, and multiply matrices.

_____ Example 2: Add matrices.

_____ Example 3: Add and subtract matrices.

_____ Example 4: Perform scalar multiplication on matrices.

_____ Example 5: Multiply matrices.

Goal 3: Determine the multiplicative identity of a matrix.

_____ Example 6: Determine identity matrices for given matrices.

HOMEWORK

_____ Textbook exercises pages 633–635 (specify) _____

_____

_____ Look Back page 635   _____ Look Beyond page 635

_____ Student Study Guide 12.8

ADDITIONAL HOMEWORK ASSIGNMENTS AND RESOURCES

_____ Practice 12.8   _____ Reteaching 12.8

_____ Student Technology Guide 12.8   _____ Lesson Quiz 12.8

Additional comments and instructions

_____

_____

_____

_____

_____

_____

_____

Make-Up Lesson Planner for Absent Students
13.1 *Theoretical Probability*

The items checked below were covered in class on _____ (date missed).

Goal 1: List or describe the sample space of an experiment.

_____ Example 1: Find sample spaces for given coin-tossing experiments.

_____ Example 2: Find favorable outcomes for given coin-tossing experiments.

Goal 2: Find the theoretical probability of a favorable outcome.

_____ Example 3: Find theoretical probabilities of favorable outcomes in number-cube experiments.

_____ Example 4: Find and compare theoretical probabilities of favorable outcomes in similar experiments.

HOMEWORK

_____ Textbook exercises pages 651–652 (specify) _____

_____

_____ Look Back page 653   _____ Look Beyond page 653

_____ Portfolio Activity page 653   _____ Student Study Guide 13.1

ADDITIONAL HOMEWORK ASSIGNMENTS AND RESOURCES

_____ Practice 13.1   _____ Reteaching 13.1

_____ Student Technology Guide 13.1   _____ Lesson Quiz 13.1

_____ Internet Connection   GO TO: go.hrw.com   KEYWORD: MA1 Blaise

Additional comments and instructions

_____
_____
_____
_____
_____
_____
_____

Make-Up Lesson Planner for Absent Students
13.2 *Counting the Elements of Sets*

The items checked below were covered in class on _____ (date missed).

Goal 1: Find the union and intersection of sets.

_____ **Activity:** Venn Diagrams for AND and OR

Goal 2: Count the elements of sets.

_____ Example 1: Count specific elements of sets that intersect.

Goal 3: Apply the Addition of Probabilities Principle.

_____ Example 2: Compare the probabilities of two specific outcomes with the probability that one outcome OR the other will occur.

_____ Example 3: Apply the concept of intersecting sets and the Addition of Probabilities Principle to find the probabilities of drawing specific cards from a full deck.

HOMEWORK

_____ Textbook exercises pages 658–660 (specify) _____

_____

_____ Look Back page 660   _____ Look Beyond page 660

_____ Student Study Guide 13.2

ADDITIONAL HOMEWORK ASSIGNMENTS AND RESOURCES

_____ Practice 13.2   _____ Reteaching 13.2

_____ Student Technology Guide 13.2 _____ Lesson Quiz 13.2

_____ Internet Connection   GO TO: go.hrw.com   KEYWORD: MA1 Fair Polyhedra

Additional comments and instructions

_____

_____

_____

_____

_____

Make-Up Lesson Planner for Absent Students
13.3 The Fundamental Counting Principle

The items checked below were covered in class on _____ (date missed).

Goal 1: Use a tree diagram to count the number of choices that can be made from sets.

_____ Example 1: Use tree diagrams to determine the total number of distinct choices that can be made from sets.

Goal 2: Use the Fundamental Counting Principle to count the number of choices that can be made from sets.

_____ **Activity:** Sets of Choices

_____ Example 2: Use the Fundamental Counting Principle to determine the total number of distinct choices that can be made from two nonoverlapping sets.

_____ Example 3: Use the Fundamental Counting Principle to determine the total number of distinct choices that can be made from three nonoverlapping sets.

_____ Example 4: Use the Fundamental Counting Principle to determine the total number of distinct choices that can be made from two nonoverlapping sets.

HOMEWORK

_____ Textbook exercises pages 664–665 (specify) _____

_____

_____ Look Back page 666   _____ Look Beyond page 666

_____ Portfolio Activity page 666   _____ Student Study Guide 13.3

ADDITIONAL HOMEWORK ASSIGNMENTS AND RESOURCES

_____ Practice 13.3   _____ Reteaching 13.3

_____ Student Technology Guide 13.3   _____ Lesson Quiz 13.3

_____ Internet Connection   GO TO: go.hrw.com   KEYWORD: MA1 Bells

_____ Internet Connection   GO TO: go.hrw.com   KEYWORD: MA1 Probability

Additional comments and instructions

_____

_____

Make-Up Lesson Planner for Absent Students
13.4 *Independent Events*

The items checked below were covered in class on _____ (date missed).

Goal: Find the probability of independent events.

_____ **Activity:** Using a Model

_____ Example 1: Find the probability of two independent events occurring in a number-cube experiment.

_____ Example 2: Find the probability of two independent events occurring by using a 10×10 square.

_____ Example 3: Find the probability of two independent events occurring in a playing-card experiment.

_____ Example 4: Find the probability of two independent events occurring by using the complement.

HOMEWORK

_____ Textbook exercises pages 671–672 (specify) _____

_____

_____ Look Back page 673   _____ Look Beyond page 673

_____ Portfolio Activity page 673   _____ Student Study Guide 13.4

ADDITIONAL HOMEWORK ASSIGNMENTS AND RESOURCES

_____ Practice 13.4   _____ Reteaching 13.4

_____ Student Technology Guide 13.4   _____ Lesson Quiz 13.4

Additional comments and instructions

_____

_____

_____

_____

_____

_____

_____

_____

Make-Up Lesson Planner for Absent Students
13.5 *Simulations*

The items checked below were covered in class on _____ (date missed).

Goal: Design and perform simulations to find experimental probabilities.

_____ **Activity:** Basketball Simulation

_____ Example 1: Create a simulation, perform a large number of trials, and record the results.

HOMEWORK

_____ Textbook exercises pages 679–680 (specify) _____

_____

_____ Look Back page 681   _____ Look Beyond page 681

_____ Portfolio Activity page 681   _____ Student Study Guide 13.5

ADDITIONAL HOMEWORK ASSIGNMENTS AND RESOURCES

_____ Practice 13.5   _____ Reteaching 13.5

_____ Student Technology Guide 13.5   _____ Lesson Quiz 13.5

_____ Internet Connection   GO TO: go.hrw.com   KEYWORD: MA1 Simulation

Additional comments and instructions

_____
_____
_____
_____
_____
_____
_____
_____
_____
_____
_____
_____

Make-Up Lesson Planner for Absent Students
14.1 *Graphing Functions and Relations*

The items checked below were covered in class on _____ (date missed).

Goal 1: Use models to understand functions and relations.

_____ **Example 1:** Draw models of relations, and determine whether they are functions.

_____ **Activity:** Vertical-Line Test

_____ **Example 2:** Determine whether graphs represent functions by using the vertical-line test.

Goal 2: Evaluate functions by using function rules.

_____ **Example 3:** Evaluate functions by using function rules.

_____ **Example 4:** Use ordered pairs of a function to draw a model of the function.

Goal 3: Identify the parent functions of some important families of functions.

_____ **Example 5:** Identify parent functions represented by graphs.

HOMEWORK

_____ Textbook exercises pages 697–699 (specify) _____

_____

_____ Look Back page 699   _____ Look Beyond page 699

_____ Student Study Guide 14.1

ADDITIONAL HOMEWORK ASSIGNMENTS AND RESOURCES

_____ Practice 14.1   _____ Reteaching 14.1

_____ Student Technology Guide 14.1   _____ Lesson Quiz 14.1

_____ Internet Connection   GO TO: go.hrw.com   KEYWORD: MA1 Weather

_____ Internet Connection   GO TO: go.hrw.com   KEYWORD: MA1 Graphs

Additional comments and instructions

_____

_____

Make-Up Lesson Planner for Absent Students
14.2 *Translations*

The items checked below were covered in class on _____ (date missed).

Goal: Describe how changes to the rule of a function correspond to the translation of its graph.

_____ **Activity:** Translations

_____ Example 1: Compare graphs that have been translated vertically with their parent functions.

_____ Example 2: Compare graphs that have been translated horizontally with their parent functions.

_____ Example 3: Identify functions that represent translations of parent graphs.

_____ Example 4: Write and graph functions representing real-world situations, and describe the relationships between the graphs.

HOMEWORK

_____ Textbook exercises pages 705–706 (specify) _____

_____

_____ Look Back page 706   _____ Look Beyond page 706

_____ Student Study Guide 14.2

ADDITIONAL HOMEWORK ASSIGNMENTS AND RESOURCES

_____ Practice 14.2   _____ Reteaching 14.2

_____ Student Technology Guide 14.2   _____ Lesson Quiz 14.2

Additional comments and instructions

_____

_____

_____

_____

_____

_____

_____

Make-Up Lesson Planner for Absent Students
14.3 Stretches and Compressions

The items checked below were covered in class on _____ (date missed).

Goal: Describe how changes to the rule of a function stretch or compress its graph.

_____ **Activity:** Vertically Stretching a Function

_____ Example 1: Graph functions and identify vertical stretches or compressions.

_____ Example 2: Sketch horizontal stretches of parent graphs, and compare their *y*-values for given *x*-values.

_____ Example 3: Identify functions that can represent stretches or compressions of parent graphs.

_____ Example 4: Graph functions, and identify horizontal stretches or compressions.

HOMEWORK

_____ Textbook exercises pages 711–713 (specify) _____

_____

_____ Look Back page 713   _____ Look Beyond page 713

_____ Student Study Guide 14.3

ADDITIONAL HOMEWORK ASSIGNMENTS AND RESOURCES

_____ Practice 14.3   _____ Reteaching 14.3

_____ Student Technology Guide 14.3   _____ Lesson Quiz 14.3

Additional comments and instructions

_____
_____
_____
_____
_____
_____
_____

Make-Up Lesson Planner for Absent Students
14.4 Reflections

The items checked below were covered in class on _____ (date missed).

Goal: Describe how a change to the rule of a function corresponds to a reflection of its graph.

_____ **Activity:** Reflection Across the *x*-Axis

_____ Example 1: Graph functions and identify reflections across the *x*- or *y*-axis.

_____ Example 2: Graph given functions, and describe the reflections of the parent graphs.

HOMEWORK

_____ Textbook exercises pages 717–718 (specify) _____

_____

_____ Look Back page 718   _____ Look Beyond page 719

_____ Portfolio Activity page 719   _____ Student Study Guide 14.4

ADDITIONAL HOMEWORK ASSIGNMENTS AND RESOURCES

_____ Practice 14.4   _____ Reteaching 14.4

_____ Student Technology Guide 14.4   _____ Lesson Quiz 14.4

_____ Internet Connection   GO TO: go.hrw.com   KEYWORD: MA1 Friezes

Additional comments and instructions

_____

_____

_____

_____

_____

_____

_____

_____

_____

Make-Up Lesson Planner for Absent Students
14.5 *Combining Transformations*

The items checked below were covered in class on _____ (date missed).

Goal: Graph functions that involve more than one transformation.

_____ **Activity:** Order of Vertical Transformations

_____ Example 1: Apply transformations in the correct order to graphs of parent functions.

HOMEWORK

_____ Textbook exercises pages 725–726 (specify) _____

_____

_____ Look Back page 726   _____ Look Beyond page 727

_____ Portfolio Activity page 727   _____ Student Study Guide 14.5

ADDITIONAL HOMEWORK ASSIGNMENTS AND RESOURCES

_____ Practice 14.5   _____ Reteaching 14.5

_____ Student Technology Guide 14.5   _____ Lesson Quiz 14.5

Additional comments and instructions

_____

_____

_____

_____

_____

_____

_____

_____

_____

_____

_____

_____